Do Your Own
DOUBLE GLAZING
& INSULATION

BY THE SAME AUTHOR

Install Your Own Central Heating

DO YOUR OWN
DOUBLE GLAZING
& INSULATION

HAROLD KING
Editor, Practical Home Building & Decorating

London
W. FOULSHAM & CO LTD
New York · Toronto · Capetown · Sydney

W. FOULSHAM & CO. LTD.
Yeovil Road, Slough, Bucks, England

ISBN 0 572 00776 0

Designed by Rosemary Harley
Computer-composed and printed in England by
Eyre & Spottiswoode Ltd., at
Grosvenor Press, Portsmouth

CONTENTS

1

COMFORT

The question of comfort in the home is assuming greater importance in most people's lives. The conception of the stoical, hardy Britisher roasting over a fire in the one warm room in the house, creeping along cold passages and, dreading going to bed, undressing in a bedroom at seemingly sub-zero temperatures is slowly becoming a thing of the past. We are gradually beginning to realize that there is no virtue in breaking the ice before we wash in the morning and that all-over house warmth can even be a pleasant thing.

Today, householders, the Government and builders alike are concerned that we have, within our homes, both warmth and fresh air. We have to decide how warm we wish to be, how to attain, and then how to retain that heat.

All forms of heating cost money. Some fuels are cheaper in particular circumstances. It is generally agreed that solid fuel and oil, particularly if one is a big consumer, are cheapest, gas more costly and electricity the most expensive. It should be pointed out that the white meter system of using off-peak electricity has narrowed the gap.

Ultimately, though, one has to face the fact that all forms of heating are expensive and yet in so many homes much of this expensively gained warmth is lost to the 'great outside' through the weak points of a house. Roofs, doors, gaps round windows, badly fitting floorboards and so on. We seem only recently to have realized that houses should be designed with full thermal insulation. One landmark in this increasing awareness was the setting up of the Parker Morris Committee in 1959 to report on the standards attained at that time and the improvement hoped for in the future. Their report *Homes for Today and Tomorrow*, which appeared in 1961, made recommendations concerning floorspace, heating and other allied topics. With the adoption of higher standards of home heating it becomes obvious that much more thought has to be given to higher standards of thermal insulation. The higher the temperature inside the house, the greater the potential heat loss and the greater the need for insulation.

Central heating, run successfully and efficiently, requires certain temperatures in various rooms maintained when the outside temperature falls to 30°F (1°C). To achieve these temperatures one must control the air changes per hour through controlled ventilation and prevent unnecessary loss of heat through ineffective insulation. Ideal temperatures vary slightly but an acceptable standard to within 5°F is given here:

Area	Temperature		Air changes per hour
Hall	60°F	16°C	2
Corridors	60°F	16°C	2
Staircases	60°F	16°C	2
Bathrooms	65°F	19°C	2
Living rooms	70°F	22°C	1-1½
Bedrooms	55°F	13°C	1-1½

The benefits of insulation may be reaped as extra warmth with the same fuel consumption or as savings in fuel with no decrease in warmth. Normally, people strike a happy medium. Insulation also helps to keep a house cool in summer and so may provide an added bonus.

When planning a central heating installation or checking to see the one you have is economical and meets your requirements you have to check the boiler sizing, its btu output, and the size of the radiators needed to keep a desired level of heat in a particular room remembering that the maintaining of the desired degree of temperature inside has a direct relationship to the temperature outside. It is necessary to calculate the building's heat losses. One has to calculate the 'u' value of the structure, that is the 'thermal transmittance coefficient'. A simple definition of a u-value is

8

the amount of heat which is lost from one side of a structure to the other, or from the inside of the house to the outside, per sq ft, per hour and per degree difference on each side. The lower the u-value the better. A high u-value means a high heat loss and a poor degree of insulation.

Below are some u-values for materials and structural parts of the typical house:

Part	Description	u-value
External walls		
Solid brick	225 mm plastered inside	0.43
Cavity wall	275 mm plastered inside	0.34-0.30
Stone	300 mm	0.50
Concrete	101 mm	0.60
Doors		0.50
Windows		
	Single-glazed	1.00
	Double-glazed	0.50
Ground floors		
	Wood boards on joists	0.30
	Thermoplastic tiles on concrete	0.20
	Wood blocks on concrete	0.15
Roofs		
	Tiles on battens	0.56
	Tiles on battens with felt	0.43
	Tiles with boards and felt	0.30
	Tiles with felt & ceiling insulation of 25 mm quilt or 51 mm loose fill	0.15
	Tiles with felt and 101 mm ceiling insulation	0.10

A systematic check on where the heat is being wasted, through gaps and cracks and other weak spots, is important. Check carefully around doors. Most doors, even well-fitting ones, have the equivalent of a 174.2 sq cm hole. This gap is around the frame where the door fits, and adds up to this surprising total. Badly fitting windows, gaping floorboards, disused chimneys and roofs are all exits for your heat. If you do block off a chimney no longer needed, it is important, unless you cap the chimney pot, to leave a ventilator, or condensation will soon attack the inside of the chimney, and show characteristic telltale brown stains on your decorations.

In the following chapters we deal with some of these insulation problems, the solution to which may help to create a warmer environment for you and your family.

The cost of insulation need not be prohibitive. In the early stages of building cost of good insulating materials vary little from those with poorer qualities. Cavity walls, having an inner skin of insulating building blocks, cost little more than ordinary bricks. They are also quicker to lay and may even therefore be cheaper in the long run. Solid floors, which are better insulants than timber floors, are actually cheaper to

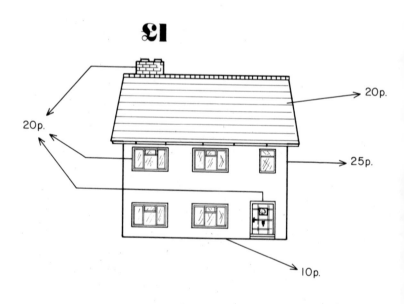

25p.	through	walls	lost
10p.	,,	floors	,,
20p.	,,	roof	,,
20p.	,,	windows, doors and flues.

lay than timber floors. While the house is actually being built it is comparatively cheap to lay a blanket of insulation under the floorboards of the house. A house of 93.0 sq metres would cost around £30 and the roof space of such a house laid with 25 mm of insulant matting would cost about the same. However, as explained later, a 51-75 mm depth of insulation is the minimum needed to obtain the fullest comfort and fuel economy.

In the average house 75% of every £ you spend is lost through the fabric of the house as a result of poor insulation. It has been estimated that of this cost, 25p is lost through walls, 20p through the roof, 20p through the windows, doors and flues and 10p through the floors. Thus, only 25p of your £ actually goes to keep you warm.

2

WALLS

When considering insulation it is logical to tackle first the surface area that wastes most heat. It is accepted that in the average house this is the wall area. The proportional amount of heat loss may go up if you live in a house that has large areas of glass, but for most of us the walls are the vulnerable areas.

First, you have to determine the kind of wall structure your house has. Most modern houses have cavity walls — consisting of two brick leaves, or one brick and one insulating block separated by an air gap of about 50 mm. The idea of building walls with these cavities is to prevent water reaching the inner walls. Unfortunately this free moving air also carries heat away and makes the wall a poor insulator.

If you live in an older house the chances are that the construction of your walls will be solid brick or stone. The only method of insulating this type of wall is to add insulant materials to the inner surfaces, the methods of which will be outlined later.

As so often happens in matters of heating and insulation, we are only just catching up with what has been fairly standard practice in Scandinavian and North American building for many years.

One method of insulating cavity walls is to fill the space between the walls. However, before one becomes too enthusiastic this is not a job to be done by the home handyman, since special machinery is used to inject this cavity filling. Several firms market a form of cavity wall filling. Although varying in small details the method is basically the same. Holes are drilled at regular intervals in the brickwork, and a core of brick — a plug — is removed. The nozzle of a high-pressure pumping machine is inserted into the hole and foam, rock wool or some other form of insulating material is pumped under pressure into the cavity. Plastic fill requires only a 25.9 mm hole whilst resin-bonded wool needs a larger one. Various firms make particular claims for their products but, basically, once injected the fill does not settle or shrink, cannot allow water to cross the filled cavity, is rot-proof, vermin-proof and non-combustible. This method requires a structure wall capable of standing up to an initial injection pressure of 18.144 kg per 6.45 sq cm. This form of insulation is also highly effective if used when the inner skin wall is constructed of insulating blocks. Once the nozzle of the pumping machine is removed, the brickwork is made good. Usually the brick plug is inserted and mortared in. Great care is taken to match the colour of the mortar and, after drying, little trace of the drilling operation will remain. The process on an average-sized detached chalet bungalow, for example, can be completed in a day and cause little dislocation.

The Heat Insulation Laboratory of the Danish Technological Institute issued in 1965 the following table on cavity walls and filling by various materials.

Cavity filling material	Thickness of wall (mm)	Cavity width (mm)	u-value
Uninsulated and unventilated cavity	275	51	0.27
Injected carbonide or plastic foam	275	63	0.14
Rockwool	275	63	0.10

If you are building an extension, you would be wise to consider cellular building blocks for the inner skin. These blocks, made of aerated concrete, are very light in weight. Technically the same method is used when building with them as with bricks. However, their insulation qualities are far greater. A 75 mm brick building block wall has the same insulating value as 325 mm of ordinary bricks.

13

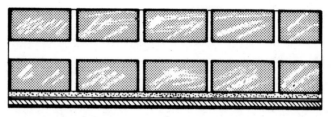

1

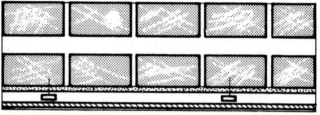

2

EXTERNAL WALL INSULATION
1. Insulation fixed to walls. Rigid insulation board, as well as polystyrene, can be fixed with adhesive, in accordance with maker's instructions. Polystyrene (13mm or ½in), u-value: 0.19: fibreboard (13mm or ½in), u-value: 0.21.
2. Board fixed to battens. By fixing insulation boards on battens 50mm x 25mm (2in x 1in) a further cavity is created. Still air is a good insulator. Fibreboard (13mm or ½in), plaster skim coated, u-value: 0.17: foil-board, plasterboard (10mm or 3/8in), unskimmed, u-value: 0.19.

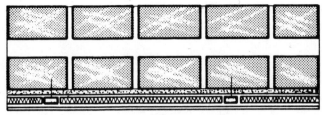

3

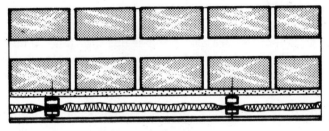

4

... /

EXTERNAL WALL INSULATION

3. Semi-rigid board (25mm or 1in), consisting of a wool material formed into semi-rigid slabs by inclusion of resin bond laid tightly between battens and covered with plasterboard, u-value: 0.14.

4. Wool insulation blankets and foils, similar to those in roof insulation, may be fixed over battens with a 2nd layer of battens for plasterboarding. Blanket (19mm or ¾in), u-value: 0.13. With reinforced foil on 9in (229mm) wall, u-value: 0.15.

If you have solid-wall construction you will have to rely on insulating the inner surfaces of your walls. Apart from the prevention of unseen heat loss you will prevent cold surfaces — which spell condensation and in some rooms death knell to your decorations.

A simple technique that will help the condensation problem is to 'hang' polystyrene wallpaper, which is very thin — barely 2 mm thick. It is supplied in rolls and is put under wallpaper or painted. Aluminium foil-backed paper is also useful for this purpose. The disadvantage of polystyrene paper is that it is quite fragile, and is easily dented. It is put up in the same way as ordinary wallpaper. It is easily cut with scissors or a trimming knife and is butt-jointed. Obviously, the greatest value of this very thin layer of insulation lies in making a warm surface and thus helping to reduce the risk of condensation.

Polystyrene tiles are also popular but they tend to be used more often on ceilings. Some people are worried about the fire risk of polystyrene. It will burn — as will most building materials if a fire starts — but it is more fire resistant if left unpainted. Mostly these tiles are of the self-extinguishing fire-resistant type. If you are putting up polystyrene tiles you should wear cotton gloves to prevent marking them.

A much more effective method of inner wall insulation — other than replastering with thermal plaster — which will help — is to clad the wall. There are several insulated, cladding boards on the market. Although they vary slightly in detail, they consist basically of a rigid polyurethene foam core, sandwiched between a variety of materials. One firm provides boards that are plastic- or paper-faced — Purlboard. It also makes a plasterboard version. Normally this type of cladding is nailed onto battens screwed to the wall. For extra insulation, a layer of mineral wool quilting can be fixed behind the battens.

Wood cladding, a natural insulator, makes a decorative insulating surface and, again, a layer of quilting laid behind the cladding gives extra insulation. It will be very expensive to insulate all inner walls but the work can be done progressively, insulating the most exposed walls first.

Another method of insulating outside walls, particularly in bedrooms, is to install fitted cupboard units.

1

2

CEILING AND WALL INSULATION
1. Thermal qualities, which raises the touch temperature and lowers the condensation factor, by Newalls Insulation and Chemical Co. Ltd.
2. Rocksil Sewn Quilt PF2 used in a loft to give thermal and acoustical insulation.

/ . . .

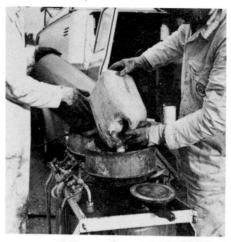

3

4

5

... /
CEILING AND WALL INSULATION
3. ICI's urea foam being fed into a machine for on-site preparation.
4. Holes drilled at intervals in brickwork are later made good.
5. Foam is inserted, under pressure, through brickwork into the cavity.
6. Cellular building blocks have thermal and acoustical properties. Blocks such as these by Thermalite Ltd. also have excellent load-bearing qualities.

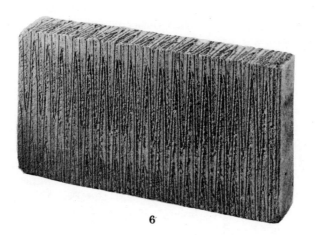

6

FLOORS

Many homes have suspended floors, that is timber-boarded floors laid on joists, with what should be a well-ventilated, dry space below. The emphasis here is on 'well-ventilated', for while ventilation is highly desirable below floor level — to help prevent wet rot and other decay — so many floorboards are badly fitting that this type of floor has a u-value estimated as high as 0.35.

The heat loss through a solid floor is a little more than half that of a suspended floor — 0.20. Floor insulation is not a particularly easy form of insulation to carry out, particularly with suspended floors. If you are prepared to make the effort, lift the floorboards and lay a quilt of insulating material or aluminium foil-backed paper across the joists to lie between them and the floorboards. You could also use two layers of 12 mm insulating fibre board. This is probably a job only to be contemplated in such detail if you have to take the floorboards up as part of the preparation for another job — such as laying central-heating pipes, renewing wiring or, possibly, for timber treatment.

An answer to the gaps between the floor and skirting boards can be to fit a piece of quadrant timber strip tightly against the skirting board. It should be fixed to the floor and not the skirting board to allow for any movement made by the floor. A traditional method is to stuff newspaper between gaps in floorboards. This is both simple and effective.

If services, gas, plumbing or wiring are below the floorboards, or run in a known section, one effective method of draught-proofing is to lay sheets of hardboard across the floorboards. Not only will this help prevent draughts rising but it will also provide an even base for lino, thermoplastic tiles, self-adhesive carpet tiles or carpet underlay. Apart from an enhanced appearance, your valuable flooring laid on a level surface will have a longer life.

Flooring coverings used may also give good insulation. One of the new foam-backed vinyl floor coverings or thick underfelt with the best-quality carpeting you can afford will help to aid insulation and add to comfort.

To help you in your choice of flooring or to estimate the likely heat loss through the floor you have, the following table may be of use.

	u-value
Wood boards on joists	0.30
Parquet or lino over floorboards on joists	0.25
Thermoplastic tiles on concrete	0.20
Wood blocks on concrete	0.15

Large gaps between boards can be remedied by inserting and tapping home slivers of wood. A proprietory cellulose filler mixed with a little PVA adhesive and stain will produce a reasonable compound. To attain a presentable surface smooth down the latter when dry and apply a uniform finish to the floor area — stain, polyurethene and so on.

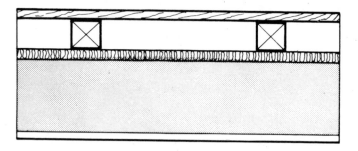

FLOATING FLOOR
7/8in (22mm) tongued-and-grooved board fixed to 2in x 2in (51mm x 51mm) battens at 16in (457mm) centres laid on 1in (25mm) quilt.

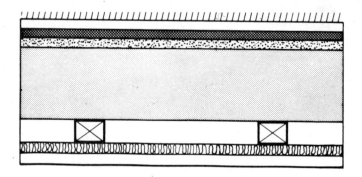

SUSPENDED CEILING
Plaster, with 1in (25mm) quilt on 2in x 1½in (51mm x 38mm) battens. Soft-floor surface.

u— = 0·16

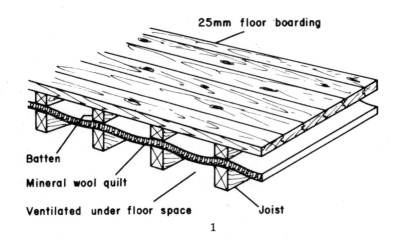

25mm floor boarding

Batten

Mineral wool quilt

Ventilated under floor space

Joist

1

1.2 FLOATING FLOOR

u— = 0·19

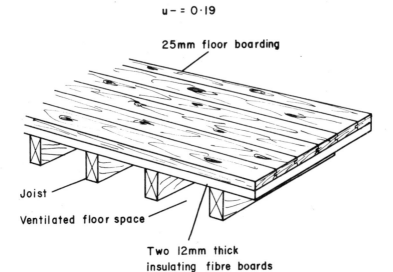

25mm floor boarding

Joist

Ventilated floor space

Two 12mm thick
insulating fibre boards

2

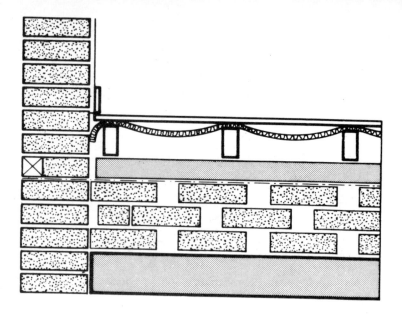

FLOOR INSULATION
1. Ground-floor blanket insulation. This should be draped over the joists and the adjacent lengths butted. The floorboards should be nailed down to compress the blanket. u-value: average: 0.15/0.18.
2. Board materials at ground floor. Heavy-duty insulating board, 1in or 25mm thick, laid over the joists. u-value: about 0.17.

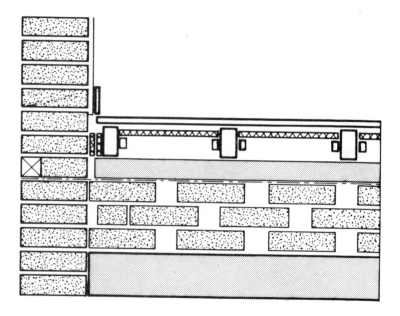

4

ROOFS

With whole-house insulation, probably the most important factor when calculating wastage of heat is the roof. The pitched roofs of modern houses usually consist of timber framing covered with clay or concrete tiles laid on battened felt. Slate, up until about forty years ago, was commonly used in Britain and many older properties have this. Concrete tiles, particularly the interlocking type, are being increasingly used because they are easy to put up. Once the first row is fixed they merely slot together. Similarly if a tile becomes dislodged or broken it is easily replaced. There is no need to despair if you have slate roofing. Although repair is a little more difficult than with tiles, slates can be replaced quite easily. Obviously, misalignment and any deterioration, cracking or crumbling of tiles or slates should be repaired, for the weak spot will allow heat to escape and damp to get in.

Rooms below ceilings are separated from the roof space by ceiling joists and, in modern homes, a ceiling consisting of plasterboard with a thin plaster skin. The u-value of this type of ceiling construction can be as high as 0.43 and a reduction to 0.10 can be effected by 51-75 mm of insulating material laid on the ceiling joists. Heat loss can be reduced by as much as three-quarters by insulating the roof space. So obviously it makes sense to afford this insulation.

The standard requirement in England and Wales is a u-value for roofs of not more than 0.25 and 0.20 in the Greater London Council area. This requirement was set out in the 1965 Building Regulations. In Scotland statutory standards are a value of 0.20 for roofs, as defined in the 1963 Building Standards (Scotland) Regulations.

The standard for England and Wales of 0.25 is bettered by many builders but it can be achieved by the use of 25 mm of mineral-fibre quilt inside the roof. In future a u-value for the roof of not more than 0.10 is likely to be the accepted standard.

An accepted second set of standards — the 'Anticipated Internal Temperature' — has also been recommended as an ideal to be aimed for. This suggests temperatures of 23°C for the living, dining and kitchen areas and 16°C elsewhere. Obviously, a warmer house means greater differences between internal and external temperatures and an even greater heat loss through the roof if insulation standards remain unchanged.

If we are to achieve Parker Morris standards of heating in our homes one of the first things to check is the roof. Many older houses have virtually no insulation and in modern houses, although things are improving, the standard of insulation may be inadequate. Heat will flow from high to low temperature areas until the temperature of both areas is equal.

Although there are a wealth of insulating materials to choose from there are only two main methods of insulating a roof space. One can either insulate the roof slope or the floor of the attic. The former is probably the more difficult. With this method one does not need to insulate separately the plumbing services for hot air will still rise from the house and the roof space will be warmed. The latter method is easier but one must then realize that the heat retention will be in the house and the tanks and pipes will have to be lagged separately for the roof space will become a cold area.

The principle of insulation is that some materials allow heat to pass through them less quickly than others. Among many insulating materials is polystyrene, now almost a household word. A by-product of oil waste, it has the distinct advantage of coming in several thicknesses ranging from 4-75 mm or even more. The disadvantage is that it is a vulnerable material and has to be put up carefully to avoid cracking and denting.

A second, very effective insulating material is mineral wool. This appears in a variety of forms from loose fill to matting to quilting and semi-rigid slabs.

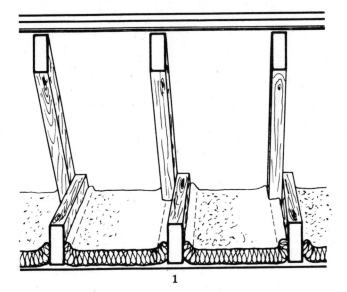

1

ROOF INSULATION
1. Glass, mineral or Rockwool blanket. This is laid between joists. Minimum recommended thickness, 1in (25mm). Up to 4in (100mm) may be desirable. u-value: 1in, 0.16; 4in, 0.08.
2. Glass, mineral, slag Rockwool or eel-grass quilt. Blanket is rolled across joists with a little slack (about 75mm or 3in). u-value: (1in or 25mm) 0.14.

/ . . .

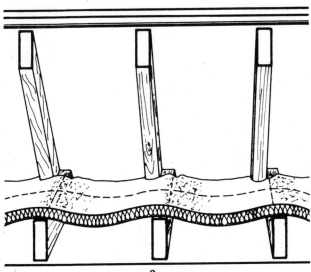

2

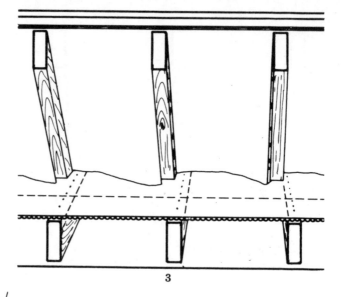

3

. . . /

ROOF INSULATION

3. Reflective aluminium foil. Combined flat and corrugated sheet, ½in or 13mm thick, laid over joists. Should be overlapped and well stapled down and tucked in at eaves. Foil is laid flat side uppermost and with slight droop. u-value: 0.21.

4. A similiar foil but laid between joists, to fit snugly. Similar u-value but with layer of crumpled foil beneath u-value is increased to 0.17.

/ . . .

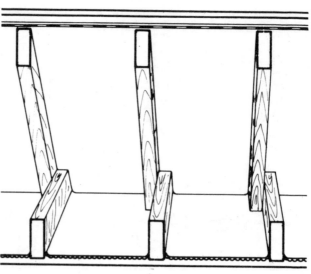

4

Many fibre building boards have good insulating properties. Most of these fibre-boards, apart from being good insulants, can also be plastered, painted or have other conventional décor finishes applied to them. The Fibre Building Board Development Organisation, of Buckingham House, Buckingham Street, London W.C.2, has a permanent show of building boards and can give full details of their properties and applications in the home.

Straw is a natural insulator as anyone who has lived in a thatched cottage will know. Cool in summer and warm in winter. Our early forebears knew a thing or two when they chose their roofing materials. One firm makes compressed straw slabs as a kit to insulate cold-water cisterns. Straw, unfortunately, is susceptible to damp and fire but can be treated to make it resistant to both fungi and fire.

Another method of lining a roof is with slabs of expanded ebonite or gypsum plasterboard. Both these surfaces can be finished-decorated and the latter is also available as loose-fill gypsum in small particles.

Like gypsum, cork can also be used as loose-fill material or as cork-board made of compressed cork granules. This may be used between two sheets of plywood or plasterboard.

As lofts are often draughty and dirty, quite a good first step is to attach bituminous-based felt or foiled-backed insulating paper to the underside of the rafters. Apart from heat conservation, this will help to keep the loft clean.

Unless one wants to use the loft space as an extra room, in view of the cost and greater difficulty of insulating the roof area it is probably better to opt for loft-floor insulation. First, deal with the plumbing services located in the loft. There are two ways of lagging pipes — with a wrap-round bandage of mineral wool or with a foam rubber sleeve. These are made in various diameters and slip round the pipe. They have the advantage of being very flexible. In awkward corners where the lagging threatens to part company with the pipe, self-adhesive tape or string can be used to secure the foam. Do not stint on lagging — always overlap the edges generously and lag right up to any termination point. It is pointless to leave weak points, especially over fittings and near cisterns.

Another, and more sophisticated form of lagging involves the use of heating elements which are wound round or strapped to the pipes. This method must be carefully installed. The pipework must be earthed and it is a good idea to have the control switch located on the upper floor of the house. It is no fun to climb up into the loft every time the weather man breathes a hint of temperatures below freezing.

A common mishap in uninsulated roofs is the freeze-up of the cold water cistern in winter conditions. Because of the expansion potential of the cistern there is little danger that it will burst, but this is not necessarily the case with pipes. For many years a hardboard outer box, filled with sawdust or insulated filling was used but more modern techniques include expanded polystyrene boarding, polystyrene slabs, which can be cut, fibre-glass wrapping, compressed straw, and so on.

For the benefit of those insulating a cold-water cistern one cannot overstate two simple precautions. The first is do not insulate beneath the cistern. If you do you are cutting the cistern off from the one source of warmth available. The very small heat loss for this area is well justified to provide just a trickle of warmth for the base of the cistern. Also, when placing insulating wraps, boards and so on around your cistern, do cut away a hole or slot to allow any vent pipe to perform its designed function. Do remember to insulate the lid or cover of the cistern.

One very popular form of loft insulation is loose fill or matting which is laid between the joists. Continental and Scandinavian countries have up to 127 mm of insulant material, but in Britain much of this is unnecessary, though at least 51 mm of insulation should be laid.

The advantages of loose fill or matting can be argued. While matting is probably easier to lay, loose fill may have better insulational properties. Air may be trapped between the granules, thus improving their thermal properties. Loose fill is

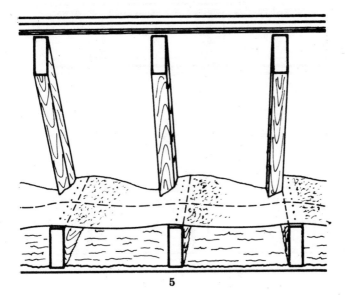

5

... /

ROOF INSULATION

5. Two separate sheets of aluminium foil, with lower sheet lightly crumpled between joists. The other sheet should be of the reinforced type, such as with a paper sandwiched between two sheets, laid with a slight droop over joists and pinned. u-value: 0.17.

6. Foil applied to roof rafters. This may be pinned, stapled or battened. u-value: 0.23; with plasterboard over, u-value: 0.18.

/ ...

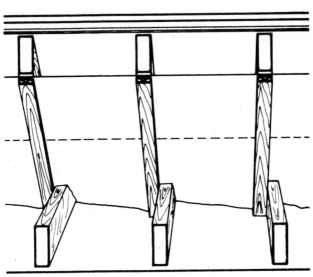

6

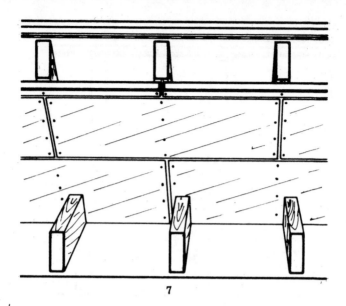

7

... /

ROOF INSULATION

7. Other materials include insulation board, cork, foil-backed plasterboard, asbestos polystyrene and urethane. All these can be applied to rafters. Boards are prefinished and in various thicknesses. Boards can also be laid on ceiling joists. Various u-values.

8. Treatment of an unfelted roof is desirable. Rain or snow may be detrimental to insulation materials. Roofing felt is available with an insulation lining and can be battened to rafters. u-value: 0.10/0.18.

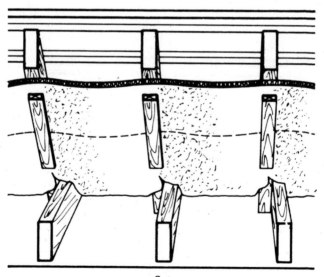

8

also an easier method if the joist widths are irregular. It is light and usually comes bagged. An average loft can be easily insulated by this method in a morning. The loose fill is simply poured between joists and levelled off to a depth of about 51 mm. A timber gauge can be easily made to spread the granules to an even depth.

One brand of loose fill mineral wool is vermin-resistant and fire- and rot-proof and comes in bags. Once laid it will remain at an even depth and not sag. It is suggested by the makers that different depths are laid for different types of house heating. Basically the more heating and the more sophisticated your heating the greater depth of insulating material required. Local heating — that is open, gas, or electric fires — requires 38 mm, while full central heating with thermostatic controls may require 75 mm. An advantage of this product is that it is non-irritant to the skin, though if used properly, this material need never come into contact with it. The fill is merely poured from the bag and tamped down.

Glass-fibre matting — which traps tiny pockets of air between the layers — is a good material to use if the joists are of even and regular width. It comes in lengths up to 500 x 400 mm. It should be pushed well under eaves — usually a cold spot. It is desirable to wear gloves when handling this material, as it may cause skin irritation. It is otherwise easy to manipulate and to cut with scissors. Glass-fibre comes in two thicknesses — one a standard matting to insulate a localized heating system and the other a super-matting for the partially or centrally heated house.

The principle is that in the first system there is less heat generated, less to rise and, therefore, less to dissipate through the roof than in a house with partial or full central heating. Of course if you do not have central heating but plan eventually to have this it might be better to look ahead and lay the thicker matting, anticipating future requirements.

The question is often asked: Does cost and initial outlay repay itself in reduced fuel bills?

The answer is that it has been calculated that householders

start making the greatest saving on their annual fuel costs after inserting 51 mm of mineral fibre quilt or equivalent insulation.

During the period 1960-9, just under £24 million could have been saved if new houses had had 51 mm of roof insulation rather than 25 mm as previously accepted.

It has been estimated the cost of insulating the roof of the average three-bedroomed 'semi' is £15. This cost should be re-couped on fuel saving within one to two years.

Once your loft is sound, your roofing secure and in good order and the loft space well insulated you are well on the way to having a well-insulated fuel-saving heat-conserving house. During a period of snowy weather look at the roofs of the other houses in your road. If the snow stays on yours longest you will know that you have done a good job. Your expensive heat is not melting the snow away; your fuel bills should be lighter and your house more comfortable.

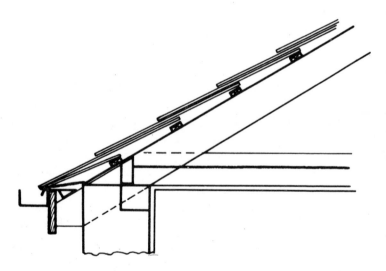

ROOF INSULATION
The underside of rafters is lined with insulation board and loose fill mineral fibre is laid between the joists. Any gaps under eaves should be sealed.

1

2

3

4

LOFT INSULATION
1. Fibreglass Supawrap is laid between the joists.
2. Granular foam, polystyrene is used as an effective between joists insulant.
3. Granular polystyrene or granular vermiculite are laid to a depth of at least 2in (51mm) between joists.
4. Stillite Products Ltd.'s pelletted mineral wool is an alternative inter-joist insulant and should be laid 2in (51mm) deep. / . . .

5

6

7

Expanded polyst

Mineral wool and
fibre

Fibre building bo

Felting

Compressed straw

Cork

Gypsum

... /
LOFT INSULATION
5. Jablite, polystyrene panels, m
by Monsanto can be cut to
between the joists.
6. A sewn quilt PF2 by Rocksil
be used for loft insulation a
sound deadening.
7. Plain quilt PF2 can be used
the same way as sewn quilt.

Spongy, plastic material normally supplied in slab form, in thicknesses varying from 6.35-75 mm or more. Will break or damage easily; has to be handled with care. Most makes are flame resistant. Comes in sheets or tiles; also supplied in very thin version in rolls.

For lagging cold-water cisterns; insulating ceilings; as underlay for wallpaper (thin version).

Comes in several forms; as loose fill in bags; in matting form or as semi-rigid slabs. Also available in quilt form. Can cause irritation or damage to bare skin. Rot proof. When in slab form, fibres are held together with bonding agent.

For insulating attic floors (loose fill and mat); fixing to joists before putting up cladding (quilt); lagging cisterns (semi-rigid). Also used in hot-water cylinder 'waistcoats' or jackets.

Comes in rigid sections and is normally made from felted wool or other vegetable fibre. There is a British Standard list of fibre boarding. Can be nailed, screwed or glued and may be painted or distempered.

Fixing to joists, rafters, battens and so on.

Supplied in rolls, it is impervious to fungal attack; jute felt, however, is susceptible under damp conditions. Felt is generally combustible.

Fixing to rafters and joists; nailing to battens before fixing cladding.

Straw is a natural insulator; is susceptible to fungus under damp conditions. Can be treated to give flame resistance. Is easy to cut; can be pinned or nailed.

For lagging cold-water cisterns; fixing to-battens, rafters, etc.

Supplied as loose fill or in slab form. Slabs can be plastered or painted. Can be used between two sheets of plywood or plasterboard.

For insulating attic floors (granulated); slab form suitable for insulating walls or attics.

This is a natural material which is processed and is becoming increasingly popular, is only intended for interior use; can be painted or plastered. Can be obtained in loose-fill form; only moderately good insulation properties.

Suitable for insulating roof slope, ceilings and so on; loose-fill for attic floors.

DRAUGHT INSULATION

Most of us are only too well aware of the discomfort of being in a room where there is a draught. Some houses are just draughty, yet, with a little forethought, much can be done to remedy the situation and save your warmth.

No house can be completely air-tight, nor would it be particularly desirable for it to be so; for our comfort and safety we need air changes but not draughts. Unfortunately, in too many houses the rate at which warm air is replaced by cold far exceeds the needs of good ventilation. Air movement within a house is caused by wind pressure and the natural tendency of warm air to rise and fill the colder spaces.

If you have an open fire of the older type as against the high-efficiency modern version, much of your heat is not radiated into the room but convected up the chimney. Why keep the birds warm, when by restricting, but not blocking, the flue one can reduce the wastage. A simple device called a chimney throat restrictor, or the use of an open fire with a built-in restrictor will achieve this end. If all flues are blocked off completely in the chimney the fire will smoke.

Here, before looking further at draught-proofing methods one must give a timely warning. Insulation does not mean no ventilation. This is often misunderstood. Most modern solid-fuel appliances and gas appliances do need ventilation and the requirements are firmly laid down in the Building Regulations. There must be a properly balanced free flow of air, not a draught.

For stopping under-door draughts there are two basic types of draught excluder. These are the threshold type which is fixed to the floor or the door frame, and the coupled type which is fixed to the door.

A popular kind of coupled excluder is one that works on the drop-bar principle. This comprises a bar of wood, metal, felt or plastic which is designed to adjust to different floor levels. A perfect seal is formed when the door is closed and when the door is opened the bar rises and rides clear of the carpet. A felt excluder is fitted to the door has the advantage of riding more easily over uneven surfaces.

Door draught excluders are designed to be unobtrusive but if you are particularly concerned about room décor an answer may be to fit an excluder which is fitted into a groove made under the door. This is obviously more of a job since a concealed excluder cannot be fitted unless the door has been taken off.

Some coupled excluders are suitable for either external or internal doors. Those for external doors are normally made of metal and contain a water-bar. If the gap is wide, draught may not be the only visitor at the base of the door and a water-bar may be essential.

Wood, plastic or metal threshold excluders are also obtainable. They are usually chamfered or shaped to prevent tripping and are fixed with panel pins, screws or adhesives. You may buy standard lengths or strips to cut and trim to your individual requirements.

As previously mentioned the equivalent of a 174.2 sq cm hole may be the grand total of the gap round what seems to be a well-fitting door. If the whole of your door is ill-fitting it will probably be necessary to use 'sprung' aluminium or bronze strips. These strips are placed around the door frame with pins placed at regular intervals and then sprung outwards so that when the door is closed the metal strip presses against the door and forms a seal. These usually have a threshold excluder on the door sill. Some metal-stripping systems are designed to be fitted by the home handyman while others, which carry a guarantee, are fitted by specialist firms.

This may seem relatively expensive but when potentially so much heat can be wasted through an ill-fitting door the cost is justified. This type of metal-sprung stripping can also be used on windows — but to fit all the windows and doors of an average size house could cost well over £100. One firm charges £12 for a modern door, but claims that its method of using a strip surround of phosphor-bronze tensioned and nailed into position on a wooden frame or riveted on a metal frame will exclude 90 per cent of draughts.

If you cannot or do not wish to contemplate this treatment for your doors a cheaper alternative is plastic or foam rubber strip which is fixed round the frame, normally with adhesive. This can either be bought packaged or by the yard.

Another type of draught-proof stripping is a foam-rubber self-adhesive stripping. The backing paper is simply peeled off and the foam is pressed into place. It is important that the surface of the window frame should be clean and dry, as dirt, damp, or grease may prevent adhesion. This type of insulating strip is cut easily with scissors.

You can even buy your draught excluder in a tube. This product comes in what looks like a large tube of toothpaste. You merely squeeze out a coil of putty-like material, and press it along the part of the door frame that comes into contact with the door. This requires a steady hand and, of course, the material must be allowed to harden before the door is shut.

If you are contemplating the installation of double glazing it is very important to check that the windows are draught-insulated.

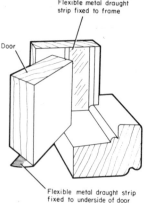

Flexible metal draught strip fixed to frame

Door

Flexible metal draught strip fixed to underside of door

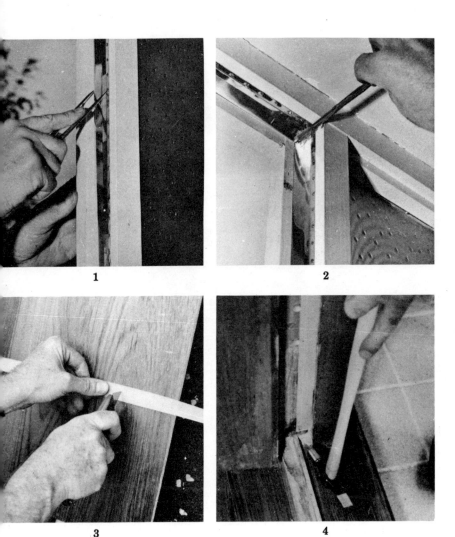

DRAUGHT INSULATION
1. A metal threshold excluder is screwed to the door frame.
2. A seal is formed by applying a mastic compound where the sections meet.
3. Tacaseal draught proofing strip is cut over size.
4. Secured with a temporary fixing clip the strip is then pinned into place.   /...

5

6

7

8

. . . /

DRAUGHT INSULATION

5. After pinning, the temporary clip is removed and the strip finger sprung to form a seal.

6. Backing paper is removed from one of the self adhesive draught proofing strips.

7. Fix strip in place round door frame then cut to exact length.

8. Simply squeezed from a tube Draught-Stop by Homecraft Product.

/ . . .

9   **10**

11   **12**
13

. . . /
DRAUGHT INSULATION
9. Metal Weatherstrips — Metaseal weather stripping for use on metal windows.
10. An excluder bar made of foam rubber or metal is fixed to the sill of the door.
11. The sill, shown in position, is made by Expandite. This provides an effective draught barrier.
12. A neat draught excluder with a self adhesive backing.
13. Neat and effective when in position.

/ . . .

14

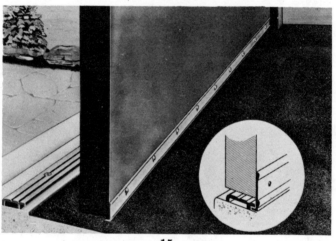

15

. . . /
DRAUGHT INSULATION

14. Floor-level draught excluder by Metal Weatherstrips, using an interior door-bottom seal.

15. Sill sealed by Metal Weatherstrips — an external aluminium seal.

6

DOUBLE GLAZING

Although double glazing has been used in Continental and Scandinavian countries for many years it has only relatively recently become popular in Britain. Many people, particularly those who would do the job themselves may be put off by the thought that the whole job might be rather 'a hit-and-miss' affair. This might be so if one went to the bother of trying to evolve one's own system but today this is just not necessary.

There are many varied double-glazing kits available. They come with almost every type of frame, plastic, metal, wood and so on. There are also several types of system, sealed units, applied sash, coupled sash and sliding sash.

Many householders consider double glazing and are deterred by the initial high cost. They want to know if it is worth while. How much will they save on heating bills? Will the home be warmer? It is a fact that a single sheet of glass is a poor insulator. In an exposed position it can have a u-value as high as 1.0 or 1.10. There is a cold 'zone' around every window — the nearer you get to the window the colder it becomes. From this we can see that heat loss through a window is higher than heat loss through a wall. It has been calculated that about 20 per cent of all heat loss in a house is via the window areas. By double glazing, this heat loss can be reduced by as much as 80 per cent, which represents about a saving of 10 per cent on the home's heat loss. So, even if the window area of your home is comparatively small, a considerable amount of heat escapes through it.

Also, many windows, particularly the older, sliding-sash type with wooden frames, are badly fitting and a source of draught as well as an escape route for your precious heat. There is little point in double glazing if heat still escapes through the cracks. Equally, there is little virtue in double

45

D

1

2

SEALED SYSTEMS
1. This factory-sealed unit by Hermaseal has good insulant qualities and the inner pane of glass cannot mist up.
2. This shows a double glass unit having 3/16in (5mm) air space made by Pilkington. It is called an Insulight Glastoglas glazing unit.

glazing if other weak points — roof, floors, walls and doors — are neglected. Therefore, combined with other insulation, double glazing certainly adds to the comfort level of the home and may reduce your fuel bills. Also the amount of thermal insulation provided by double glazing depends upon the gap between the two sheets of glass. A double unit with a 5 mm gap has a u-value of 0.60, while one having a gap of 12 mm could have a u-value as low as 0.52.

A point worth mentioning here is that of acoustical insulation. Double-glazing units do not have very high acoustical insulation properties. While a normal domestic double-glazing unit may have a gap varying between 5-12 mm, to have effective sound insulation the sheets of glass should be at least 101 mm apart. As will be readily appreciated this is not possible with most types of domestic window.

Although pre-sealed units have many advantages, cost makes the majority of people turn to one of the many self-assembly kits. A very popular method is that of applied sashes which are fitted to the surround or rebate of the window and not to the frame. If your window has an external rebate there is no reason at all why your double glazing cannot be fitted to the outside of the window. With most applied sashes, a rebate of 75 mm is necessary. 32 oz glass is usually used for the second glazing sheet.

Some factory-sealed units are supplied with a frame; while others need to be fitted to a frame which you provide. Here measurement is most important. If the unit is to be glazed into an existing wood or metal frame, a rebate of at least 12.7 mm will be required. This is because most sealed units have an air-space of 5 mm. Some units with smaller air-spaces are available for use with smaller rebated frames, but it is up to you to check which one you require before ordering. The old carpenter's maxim 'measure twice cut once' must apply here. There is nothing of less use than a custom-made, double-glazed unit a fraction of an inch or so out. If you are using an existing frame, check that it is sound and well constructed, as sealed units weigh at least twice as much as a single pane of glass. Beading is the most satisfactory method of fixing the units, but provided the unit is set into a non-hardening compound, putty fronting can be used.

Applied-sash and coupled-sash units are probably the two most popular forms of double glazing. The former is fixed to the rebate or other surround of a window, while the latter is fixed to the window frame. Sliding-sash units are also a form of applied sash in that they are fixed to the rebate of a window. You may think that a disadvantage of sliding-sash windows is that you must have a true surface if the windows are to slide perfectly. This is not necessarily so, for most firms supply packing pieces to ensure a true framework. Effective thermal insulation is achieved if the gap between the original window and the new unit is 25 mm. Acoustical insulation will be improved if the gap is increased by up to 200 mm but, of course, you need a deep rebate to do this.

You may favour a coupled-sash method of double glazing. This is a simple-to-install and effective form of double glazing. It involves fixing a glazed panel, which can be removed easily in the summer, to the existing window. Normally, the frame of the glazing unit should be 19 mm less all round than the dimensions of the window opening. The glass should be 8 mm less than the dimensions of the frame. The main frame channel is cut to size, the glazing strips fitted round the glass, and the glass is laid flat. Then the components supplied with a particular system, usually four main channels and four corner hinges, are put roughly in position. The locking nuts and screws can then be loosely fixed to the corner units. Neoprene sealing strips are cut to length and mitred at the corners, then inserted into the channels. The corner hinges are fitted to the main-frame channel, then the glass is fitted to the main frame. The sliding foot of the window-stay usually provided is fitted into the groove in the bottom channel. Then the whole unit can be tightened up. The desired position of the completed frame is ascertained by holding the unit to the window opening and marking the position of the hinges. The unit is then fixed

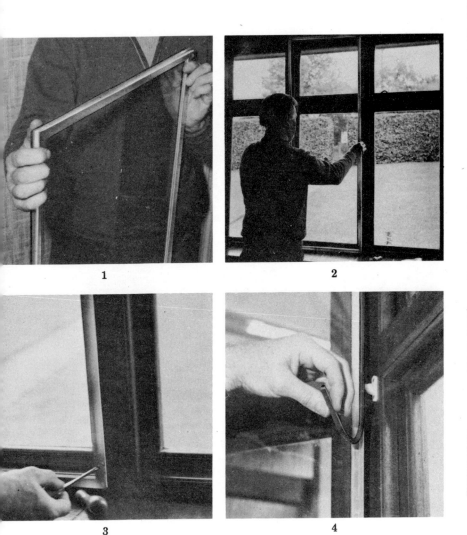

APPLIED SASH

1. This shows the assembled section of the Cosyframe de Luxe slot-together frame.
2. The assembled unit is offered to the window frame.
3. Screw fixing, directly to the frame.
4. The joint between the unit and window frame sealed with a Neoprene seal.

/ . . .

5

6

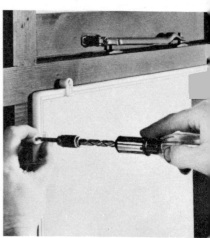

7

. . .'/
APPLIED SASH

5. A system consisting of snap-fitted PVC which simply fits round the existing window frame. Made by Oxford Double Glazing and called 'Grippa-Frame'.

6. The assembly of the main frame of a Tebrax Ltd.'s sliding panel double-glazing system.

7. Quick-Fix System is installed with plastic channelling and swivel clips, it comes complete with screws and is made by F. P. A. Finnegan Ltd.

/ . . .

8

9

10

APPLIED SASH
8. Mosanto's acetate sheet fixed to the surround with double sided tape.
9. The sheeting is placed over the adhesive tape and all wrinkles are eased out.
10. Trim sheeting to exact size using knife against a straight edge.

into place. Obviously, details of particular makes of coupled-sash vary in minor details but fitting is neither a long nor a very taxing task.

Another method of double glazing that ensures a more-or-less draught-proof window involves the use of plastic framing and beading. Coils of plastic framing with a self-adhesive base are fixed directly to the window surround. The framing is mitred at the corners. The glass, 24 oz, is pressed into the glazing channel. Finally, the plastic beading is positioned and rolled into place.

Probably the most effective form of double glazing is the sealed unit. Unlike other forms of double glazing, this involves removing the existing glass. As existing frames are not always sturdy enough to take the weight of the new units, new frames may be needed and cost can mount up alarmingly.

However, it is worth considering this system carefully. The advantages are quite profound. First, they are neat; there are only two surfaces to clean and the unit looks just like a single sheet of glass. Many people have difficulty in telling the difference between sealed double-glazed units and a single sheet of glass. Condensation is not a problem since the units are hermetically sealed. There are no movable parts and once installed, they can be forgotten about.

It is of course not always necessary to have new frames. Sometimes the existing frames can be adapted and there are a number of sealed units made with stepped rebates. Thus, one part of the unit fits into the rebates and the other overlaps it.

Many glazing firms now make double glazing easy for you. They come, measure, order and do the job. You write the cheque. Obviously this is quite expensive but you do know that the entire job is their responsibility. If, however, you like a challenge and can use your ruler accurately — although some firms will even measure up for you — there is no reason at all why you should not order these units and install them yourself. Sealed units are available in over a hundred standard sizes so you should have little problem in ordering for your requirements. However, if you do want the 101th size most firms will make a unit for you though a 'one-off' order will cost a little more and take longer to supply.

While double glazing is well worth considering, it will not immediately and dramatically cut fuel bills, but it will conserve warmth and help eliminate draughts and cold zones in a room. Essentially, though, if money is to be spent on heat conservation, you should work on descending order of priorities and after a well-insulated roof, elimination of draughts and wall insulation, the money spent on double glazing will be a sound investment for the future comfort and well-being of your family.

While concentrating on double glazing one must not forget that there are extra steps that can be taken to reduce heat loss through window areas. Heavy curtains lined with foil-backed lining material, will add to the comfort and warmth of the room.

Metric modules are being introduced for glass, and the conversions into approximate metric equivalents are as follows:

Imperial	Metric
18 oz	2 mm
24 oz	3 mm
32 oz	4 mm
3/16 in	5 mm
7/32 in	5.5 mm
1/4 in	6 mm

1

2

3

4

5

6

7

8

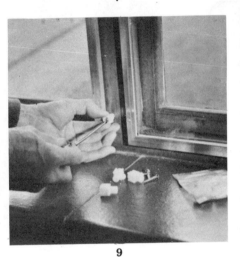

9

COUPLED, OR HINGED, SASH
1. A door hinge secured in position on a Crittal-Hope Warmlife door.
2. Checking for correct hanging. Offering up door panel.
3. Tacking the perimeter weatherstrip to the timber door frame with brass pins.
4. After marking position of side light hinges they are screwed home.
5. Slotting into position a sidelight on hinges.
6. A tough rubber seal to exclude draughts round doors.
7. Sidelight and sliding door units — complete, made by Warmlife.
8. Positioning Cosyframe de Luxe bottom-hinge pivot prior to fixing.
9. Unit held in place by unobtrusive and neat sealing catches.   / . . .

10

11

12

13

. . . /

COUPLED, OR HINGED, SASH

10. Four sealing catches are needed for larger units.

11. A cap hides the screwheads once the sealing catches are secure.

12. The completed, vinyl framed unit in position. It is not necessary to mitre corners.

13. An assembly of fixed and hinged Cosyframe sashes. The centre sash is fixed.

/ . . .

14

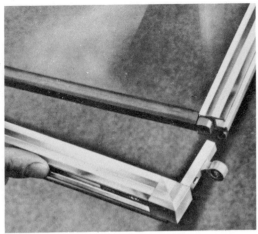

15

16

17

. . . /

COUPLED, OR HINGED, SASH

14. 15. A ready-made unit, Sheerglaze by Curico Ltd. Ready-made nylon corner pieces avoid the necessity to mitre corners.

16. A hinged section supplied and fitted by the Hermaseal Division of Rentokil Laboratories Ltd.

17. Autoplan Home Improvements supply a lightweight aluminium frame for fixed or hinged units.

/ . . .

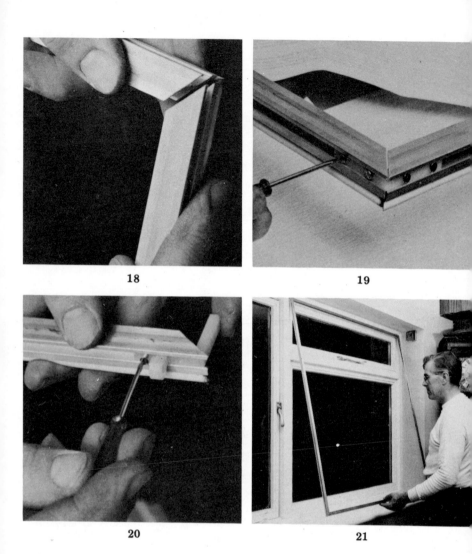

18

19

20

21

. . . /

COUPLED, OR HINGED, SASH

18. A mitre box supplied by manufacturer can be used to cut Curico aluminium units.

19. The corner pieces are screwed into the frame channel.

20. Sections held together by nylon corner pieces.

21. The finished unit hinged at the window head. The unit can also be side hung.

SLIDING SASH
1. Checking that the sill track lines up precisely with the top track the Warmlife Mk III is screwed into position.
2. Marking before screwing the top track into place.
3. The top track is screwed into place. Plugs are made in window reveal to take necessary screw fittings
4. Slider frame with side members similarly fixed.   / . . .

5

6

7

8

. . . /
SLIDING SASH
5. When the frame is squarely in position the sliding panel is inserted carefully.
6. The panel slides easily on a slim track of aluminium.
7. Adjacent panels are held firmly together with swing-over catches.
8. Everest made to measure, double-glaze sliding windows by Home Insulation Ltd.
9. An aluminium framed double glaze unit by Servowarm called Total Fit.
10. The satin-aluminium frame by Cosyframe de Luxe.
11. Automatic locking device on sliding window units by Everest provide an extra, a burglar deterrent.
12. Hermaseal Division of Rentokil Laboratories Ltd.'s custom built sliding sash unit.
13. An easily lifted out for cleaning unit, the Sealomatic unit by Weatherseal Windows Ltd.

9

10

11

12

13

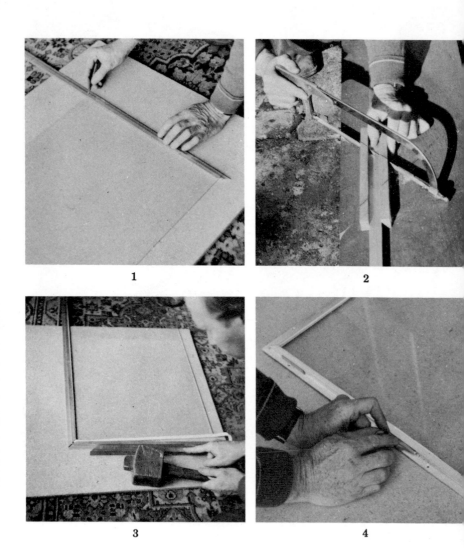

EXTERIOR DOUBLE GLAZING
1. Glass placed on the self-measuring line is marked on the aluminium strip.
2. Cutting to size the marked length of extruded frame using a mitre block.
3. Position the PVC channelling on the glass then tap the frame home.
4. The ventilation holes were drilled and the ventilator slide is positioned. / ...

5   **6**

7

. . . /
EXTERIOR DOUBLE GLAZING
5. Cutting to length the self-adhesive mastic seal for the inner frame.
6. Make sure to run each strip of seal carefully along the aluminium frame.
7. Make sure both sides of the glass are absolutely clean before they are fixed.   / . . .

8

9

10

... /
EXTERIOR DOUBLE GLAZING
8. The glazed panel is lifted into position on the outside frame before fixing.
9. The panel is secured to the exterior frame with the special screws provided.
10. The same procedure is followed for further glazed panels.

LAGGING AGAINST HEAT LOSS

When installing a central-heating system we are concerned with obtaining the maximum warmth for our money. In a central-heating system some surfaces are meant to contribute to the heating of the house and others would be better covered with insulating materials. This applies mainly to boilers, pipework and hot-water cylinders.

In the average, smallish, home the boiler is usually located in the kitchen. It will usually be covered with an enamelled steel outer case — partly to enclose working parts, and partly for cleanliness, attractiveness and insulation. One does not normally try to insulate the boiler to any greater degree as the heat emission goes towards the general heating of the kitchen.

All pipework should also be insulated. As mentioned in the chapter on roofing, pipes can be lagged either with bandaging or a foam sleeve which is sold in sizes suitable for different pipe diameters. It is very important to insulate small-bore pipes under floors and in roof spaces. It might be argued that the pipes will help to warm the underfloor space and keep it dry. But in fact the heat loss is so great that it would pay to increase the radiator size slightly and lag the pipes. Your underfloor problems can best be solved by adequate ventilation. This is achieved by keeping air vents clear.

Lagging of 25-32 mm is recommended for hot-water cylinders though if an immersion heater is installed the insulation should be at least 75 mm thick. There are many types of effective lagging. Normally a hot-water cylinder is lagged with a specially marketed jacket or 'waistcoat'. These normally consist of segments filled with 25 mm of mineral wool and fitted with bands and clips. Hot-water cylinders are lagged to retain the heat. If the use of a jacket is difficult the cylinder may be surrounded by a box made of plywood or some similar material and loose-fill insulating material, such as mineral wool or polystyrene, packed to a depth of 51-75 mm. Even with this amount of insulation the cylinder will emit enough heat to maintain an efficient clothes airing temperature.

Another way of saving heat loss within your power if you are installing your own central heating is to plan the layout of your domestic hot-water system as compactly as possible. The minimum run of pipework will reduce the surface areas from which heat can be lost.

Pipework to and from the cylinder should be lagged if the cylinder is heated by the boiler, and if the combined length of these pipes exceeds 6.10 metres — especially if the boiler is to be used in summer.

Apart from any saving on fuel bills one should have better hot-water services. The heat-loss for a compactly laid out, properly insulated central-heating or domestic hot-water system may be as much as one third less than that of an uninsulated straggling one.

A good standard of insulation which reduces the heat losses from the building not only saves on fuel costs, but since the boiler and radiators can be smaller it also reduces the initial cost of the installation.

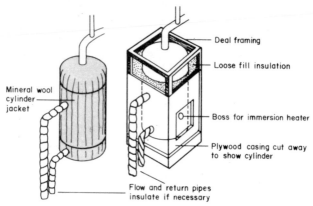

Deal framing

Loose fill insulation

Mineral wool cylinder jacket

Boss for immersion heater

Plywood casing cut away to show cylinder

Flow and return pipes insulate if necessary

CYLINDER LAGGING
Two methods of insulating a hot water cylinder.

1

2

3

4

CISTERN AND CYLINDER LAGGING
1. This shows a Rosslite polystyrene kit. The simple fixing clip is holding together the sides of the kit.
2. The lagging is complete when all the clips are in position.
3. The polystyrene cuts easily to admit pipework.
4. This shows drilling to remove a section to admit pipework — in a Stramit, lagging kit.

/ . . .

5

6

7

8

. . . /

CISTERN AND CYLINDER LAGGING

5. Once drilled the section is easily cut out.

6. Pipework in position, the cut out section is replaced, the walls positioned and then screwed.

7. Showing the lid fitted and the complete unit in place.

8. A simple cistern lagging using mineral-fibre blanket secured with string.

9. This cold water cistern is fitted with a fibreglass tank-wrap. The loft and all pipework has been insulated with fibreglass pipe-wrap. Fibreglass superwrap is laid between the joists.

10. Foamcote sectioned polyurethane foam is shown being fitted to a hot-water cylinder.

11. Band fixings, supplied, are used to secure sections.

12. A bandage type lagging used to insulate pipes.

9

10

11

12

CONDENSATION

Britain has a residually damp climate, and in winter the air contains a high proportion of invisible moisture.

Condensation is formed when warm, damp air reaches a cold surface. The condensation problem in a house is caused by what is called the 'dew point', the temperature at which moisture in the air condenses and becomes water. The water in warm air meets the wall and encounters a cold condensing barrier. This condensation can be reduced or prevented by surface insulation. If the surface is warm the moisture-laden air will meet a warm surface and will not condense. Otherwise, water in the air condenses from vapour and runs down the walls in the familiar way.

Great damage can be caused to a house both structurally and to the décor by condensation. There is nothing so depressing and, in some cases, unhealthy, as the atmosphere that goes with streaming walls and windows. In winter, particularly, we shut windows in an attempt to retain heat but in rooms that have a high moisture output — bathrooms and kitchens — there must be adequate and regular ventilation, either by window or extractor fan.

Our ancestors in earlier times lived in houses built with thick walls, which were porous enough to absorb all the condensation. Open fires roaring below big chimneys provided good ventilation. We, today, can hope to counteract condensation in several ways. Probably the easiest is to step up the heating and keep the temperature in the house at a high and constant level. If adequate ventilation is provided all should be well — except when the quarterly bills arrive. Obviously most of us are trying to keep an even balance between high bills and a comfortable, condensation-free environment.

Unfortunately, even some of the fuels we burn add to our

moisture problems. Gas and oil heaters, each burning with a naked flame, will give out water on combustion. A gas appliance will give out six pints of water for every therm of heat and an oil heater will give out five pints of water for every therm — roughly pint for pint. Even we humans give out moisture continually in our breathing. Bedrooms always require a measure of ventilation.

Probably the most important things to tackle first are the condensation weak spots — places where there is a high vapour output. Try to ensure that there is heating — even if only at a low level — in kitchens and bathrooms and in rooms facing north. Ensure adequate ventilation at all times — especially when a bath is being run or food is cooking. A simple tip — if you run a small amount of cold water into the bath first, nothing like the amount of steam will form when the hot water is drawn. Try to use absorbent soft furnishings, line ceilings with polystyrene tiles or sheeting which is warm to the touch and therefore prevents condensation forming,

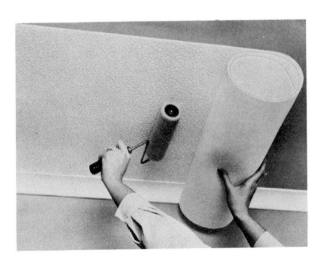

A roll-on ceiling veneer such as Warmalux has both thermal and decorative qualities.

and insulate cold spots such as concrete lintels over doors and windows with thin polystyrene. Use absorbent wall finishes. A thin polystyrene lining paper hung underneath the conventional wallpaper will again warm the wall surface.

Cork, available in tiles or sheeting makes another absorbent surface covering and can look especially attractive in a bathroom.

Extractor fans which usually have an external draught shield, are an effective and scientific way of keeping your home moisture level at an acceptable point.

Finally, the message is keep rooms warm and well ventilated and you will go a long way towards eradicating the problem of condensation in your home.

9

HUMIDIFICATION

While excessive condensation is both uncomfortable and undesirable in the home, so is the reverse. There can be the problem of too-dry air.

In some respects, heating and insulation are a contradiction in terms. On the one hand we should attempt to conserve expensive heat; on the other there should be adequate ventilation to ensure a satisfactory number of air changes, to expel stale and damp air, and to ensure air for satisfactory combustion of heating appliances which depend on the internal air for combustion.

It is all primarily a question of balance, and it is wise to make provision for combating dry air where central-heating systems are in operation.

Many people, while agreeing that warmth, ease and comfort of central heating is desirable, refuse to consider it. One of the main objections is that centrally-heated atmospheres are so dry. While many of us can tolerate moderate changes in our environment most of us tend to react to a too-dry atmosphere with headaches, sinus troubles, lethargy and, sometimes, disturbed breathing. Our systems become used to a certain humidity in the environment and if this humidity is out of balance — either too high or too low we start to suffer.

We live in a country that has a temperate climate. The normal air changes entering our homes will have a high moisture content and it is only when the outside temperature drops drastically over a long period that we find a drop in the moisture content of the air.

Air humidity is measured as relative humidity (rH). This is defined as the percentage of moisture or water vapour in the air by weight at any given temperature to the weight of water vapour required to saturate the air at the same temperature.

The following table gives examples:

About ¾ pint at 32°F =	100% rH =	very high humidity
50°F =	50% rH =	medium correct humidity
70°F =	25% rH =	very low humidity

This may sound complicated but what it really means is that when this figure of relative humidity is too low we react to what we loosely term a stuffy atmosphere. Conversely, if the humidity level becomes too high we suffer in what becomes a damp, 'hot-house' atmosphere.

Humans require a certain degree of humidity for their comfort. This varies from person to person. A recommended level of 50% relative humidity is suggested for rooms of 68°F (20°C) though it may vary a degree or two either way.

Therefore we have to seek a solution to balance our living environment — for, apart from our comfort, too dry an atmosphere can do untold damage to woodwork, furniture, pianos, pictures and soft furnishings. It sounds slightly dramatic but one could imagine a 'horror type' slow-motion film of your living-room showing plaster cracking, window frames warping, joints on furniture pulling apart, veneer cracking on the picture, which just might be very valuable and moisture being drawn out of soft furnishings and plants, pianos going out of tune, pages of books curling and so on. It sounds rather absurd — but this is just what is happening invisibly all the time in a dry atmosphere — we see only the ultimate results.

Water too is drawn from us by the dry air, our hair becomes uncontrollable, our eyes irritate, the skin becomes parched — all very unpleasant.

We could solve this problem by opening windows. In our climate — particularly on damp winter days we would soon have all the humidity we require and a gigantic drop in temperature. Obviously this is not the answer. We have to find another way of introducing moisture, in a controlled way, into the atmosphere.

This humidifier holds 70 fluid ounces of water which is evaporated in twenty-four hours at average room temperatures. It is the Ergo Model 2.

Preventive humidification means preventing the dry air from extracting moisture from the surroundings. The best way to deal with this is by using electric humidification units or hang-on-radiator humidifiers. The latter may be made of rust-free metal, china or plastic. Basically they are water containers which hook onto radiators. In the most basic type the water merely evaporates, as the radiator heats. Other humidifiers have a cotton pad which absorbs the water. As heat builds up water evaporates through the pad into the atmosphere. There is little maintenance. The inexpensive pad may need renewing perhaps once a year as various mineral salts deposited on the surface prevent effective evaporation. A good humidifier will evaporate up to five pints of water a day, so needs topping up frequently.

An electric fan-assisted humidifying unit working on the

same principle but with dispersal of water assisted by a fan may be useful as a whole-area unit on a ground or upper floor.

Water — converted into micro-atomized particles — can combine with the dry air and be ejected into the dry atmosphere by an atomization humidifier. These are quiet in operation and very cheap to run when you consider their benefits. This type of unit need only run twice a day to freshen the air. In fact if it were run for 24 hours a day, seven days a week it would still cost only in the region of 10p per week.

SOUND INSULATION

So far, in talking of insulation, we have meant insulation of our homes to save money and to give a greater degree of comfort and warmth to our environment.

There is, however, another form of insulation that is of profound importance — so important that we may not realize just how important yet — the insulation of human beings from sound and its abuse — noise. Sound in its basic form is the vibration of air sensed by the human ear. Noise is sound which is transmitted at such a high level as to be objectionable.

Unless we happen to live in an isolated rural area or on a desert island we are continually bombarded with noise. Try a simple test. Close your eyes, listen for one minute — and then make a note of all the sounds you can remember hearing. You will be amazed. Obviously some will hear more than others, depending upon where one lives but a typical list for a semi-town area might run like this: traffic, car hooters, aircraft, lawn mower, transistor radio playing, door slamming, people's voices, footsteps, children crying and so on almost indefinitely.

For most of the time, we have a built-in safety valve — for our sanity's sake, we only register these noises subconsciously — but occasionally and increasingly in these days of high-density living, we are becoming aware of noise and the need to withdraw from it into a quieter atmosphere. The problem of noise has increased during the last two decades, mainly due to the increase in traffic flow and the swing from traditional heavy building materials to lighter-prefabricated construction.

Audible sound has to be transmitted through the medium of air. It will not travel through a vacuum. Sound radiates out from the source in much the same way as the ripples that are formed when a stone is thrown into a pool of water.

The level of sound is measured in decibels (dB). The human ear becomes used to a particular range of decibels from around 8 dB, the sound made by a match being struck, to 100 dB, the noise of a jet aircraft.

The human ear hears between a range of 0 dB, the threshold of audibility, to 130 dB, at which level a sound will be felt as well as heard. A noise having a dB level of 120-130 can actually damage the hearing mechanism.

Two types of noise affect us in our homes; impact noise, which is transmitted via the building structure, and airborne noise, transmitted via the atmosphere.

High-rise living, with many people living closely together often in buildings with thin partition walls and noisy corridors, echoes, footsteps overhead, noisy children, noisy plumbing and the like — all this can add up to what can become a real anxiety state. All this may sound rather alarming but you should begin to realize that the prevention of sound transference is an important social necessity as much as a personal comfort medium.

Traffic is probably one of the most troublesome sources of noise. Travelling twenty yards from a house a heavy vehicle will batter the house wall with up to 90dB of noise. Much of this sound — up to 50 dB — is absorbed by the wall itself. Unfortunately, many present-day building materials are very light and while providing economical and easy building methods, may not provide adequate sound insulation. Noise travels through the structure of a building. The more solid the structure the less amount of noise gets through.

Noise travels in direct routes to a building and it is also reflected by hard surfaces, such as pavements and walls. Little can be done to increase the acoustical properties of the basic wall structure as this involves adding 'weight' to them. But the home improver can do a great deal by adding certain wall finishes. Some, such as polystyrene, are designed to go under conventional wallpapers and others, like cork and wood-expanded plastic and polystyrene veneers, are decorative as well as having an insulant value.

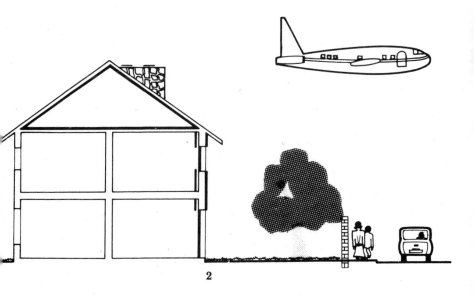

1

SOUND INSULATION

1. Exterior noise travels in a direct line and is also reflected on to a building from hard surfaces, such as pavements.

2. All the external elements of a building are equally important in insulating it against noise. Landscaping is also a factor in insulation — trees, plants and shrubs can absorb and deflect sound.

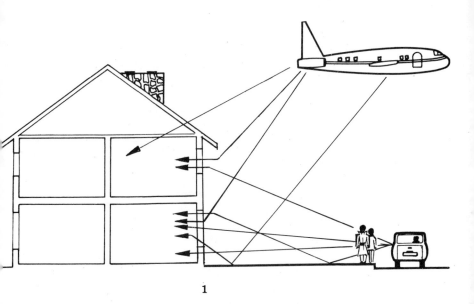

2

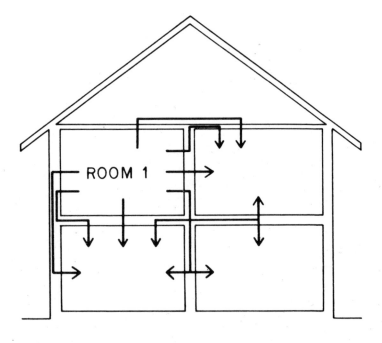

SOUND INSULATION

Noise originated in one room can be transmitted to others, as this diagram shows. The best way of reducing the transmission of sound is to reduce it in the originating room, so that it reaches others at an acceptable level.

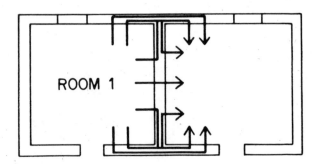

Apart from helping to conserve heat these materials absorb sound as it enters the room and help to prevent its being transmitted further round the house.

A second skin to the inner wall will help. A simple method is to fix battens to the wall and fix either a decorative or fibre-board cladding.

If you leave a gap of at least 101 mm between the outer and new inner wall you should achieve greatly improved sound insulation, as well as improved thermal insulation.

One method of reducing airborne sound is to use partition-board — sandwich panels — composed of plaster-board or hardboard, faced on each side with polystyrene. These sandwich boards are light, easy to handle and to cut. They have good acoustical properties when used with plasterboard. A greater degree of sound absorption is possible by using two or more partition boards separated by a layer of glass-fibre sound-deadening quilt. Two 51 mm partition boards with 12 mm of glass-fibre quilt separating them can give a sound reduction of up to 42 dB.

Ordinary wood studding and battens can be used to which timber cladding, cork panels, coated chipboard and so on can be added. If a gap of at least 101 mm is left between the two walls, the cavity can be filled with granulated cork, mineral fibre, polystyrene or granulated vermiculite.

In new buildings much can be done to make a quieter environment for the new owner. Much of this responsibility lies at the door of the planners. Many have taken up the challenge and, increasingly, living units are being separated from vehicular traffic. It seems pointless to build a new estate near a proposed new road or airport, if in the planning stage this can be avoided. Obviously, this is a little idealistic but much distress could be saved by careful planning.

Bedrooms and living rooms are ideally situated away from noise sources. Trees, shrubs, and solid walls of brick or stone can help to deflect noise. They act as a barrier and sound bounces back off them.

Use of acoustically good building materials, cavity walls

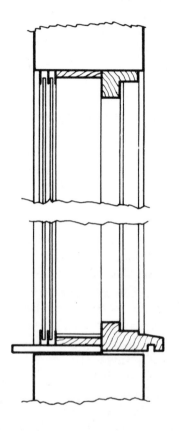

SOUND INSULATION
Double windows used to improve sound insulation. An air space of at least 4in (102mm) is necessary. Position sliding units on the inner edge and line reveals with insulation board.

filled with plastic foam or inorganic loose fill, aerated, cellular building blocks, good roof insulation, double glazing, true-fitting door and window furniture and well-lagged plumbing, are all ideals to be strived for.

No doubt some of this will increase the cost of new houses but not prohibitively so and many people, embarking on what will probably be the most important transaction of their lives, will choose to pay a little more for the bliss of quiet in their homes.

Windows can be insulated against noise with varying degrees of success. To insulate satisfactorily a gap of a minimum of 101 mm and preferably more — up to 225 mm — must be left between the two surfaces of the glass. Even just replacing the existing glass with heavier glass will help. Lining the window surround with absorbent material, such as polystyrene veneer, will also help to reduce noise. Insulant materials absorb the noise.

Roofs also need attention in the battle against noise. In much the same way as you insulate your roof against heat loss, fibre-glass quilting, boarding, loose fill and the like will help to put up a noise barrier.

A more skilled job is that of 'pugging'. Here, you lay 51 mm of sand between the joists. You must take care that your ceiling and joists are up to bearing the additional weight. Special 'pugging' boards of 12 mm chipboard can be laid between the joists to hold the sand.

Should your problem be in the form of particularly noisy neighbours, and all pleas on the grounds of neighbourliness and tolerance fail, it might be worth erecting a dry-wall partition on the party wall side. This will not reduce the sound as much as a brick or building blockwall but it will cut down the problem.

In this case the sound may be both impact and airborne. If you do not want to go to the great expense and upheaval of building a solid wall of brick or cellular building block — which must be at least 101 mm — and butted up to or tied into the main wall — a less expensive method is to construct a dry-wall partition. This dry partition is constructed as an isolated unit 200-250 mm from the main wall. There are proprietory forms of dry-walling available commercially. Basically, sound is most effectively reduced by insulation at source.

Bookshelves, preferably floor to ceiling, placed along 'noisy walls' will also help to reduce the transmittance of noise. Lining with absorbent materials, as mentioned earlier, and even with such decorative materials as hessian or felt will help to absorb noise.

Check that there are no gaps between the joists along the party wall — much of your neighbour's TV programme may travel in this way.

If you have ever walked round a bare house you will probably have noticed the echo effect. How every sound seems so much louder and distorted as it bounces off hard unsoftened surfaces, windows, walls, ceilings, floors. Once the soft furnishings are in a room much of this noise is absorbed. Heavy curtains hung at the window will keep out more than the weather.

Impact noise is most commonly caused by footsteps on suspended floors and stairs. Foam-, felt-, or cork-cushioned flooring helps to reduce impact noise and add to comfort. As thick a carpet with as good an underlay as you can afford also helps to cut noise.

Another way of reducing impact sound transmittance is with a 'floating' floor. One product suitable for this is Rockwool which is available in resilient rolls, faced on both sides with waterproof paper and stitched through.

This is unrolled across the line of the joists. Battens are then set across these and half nailed down. The fixing should not be full or in any way permanent, since if there were direct nailing between floor and joists, the effectiveness of insulation would be reduced. The nails merely hold battens temporarily in place and are later extracted.

Chipboard flooring, for example, can be set across this 'floating' floor structure. This can be screwed or nailed to the 'floating' battens.

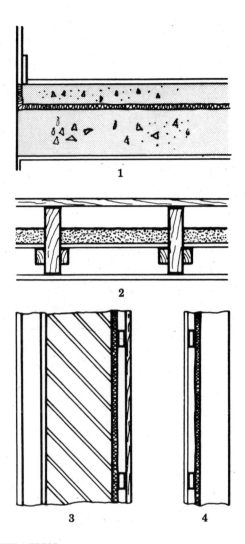

SOUND INSULATION

1. A 1½in (38mm) floating floor of sand and cement screed laid on a 1in (25mm) quilt.

2. Dry sand pugging laid on ½in (13mm) chipboard. This increases the weight of the floor by 2in (51mm), so check your joists.

3. Wall insulation consisting of an inner skin of brick built against an existing wall and mortared to it. It can be expensive as foundations may be needed. The outer face is clad with weatherboard fixed on battens. There is an extra insulant layer of quilt.

4. Lining of ¾in (19mm) plasterboard added to a partition.

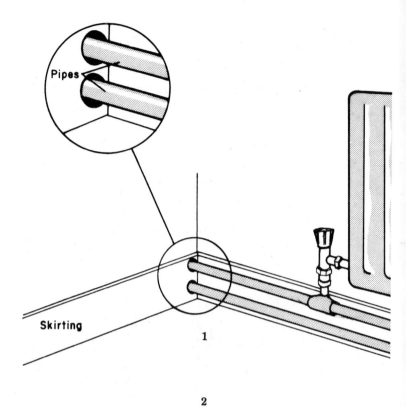

Pipes

Skirting

1

2

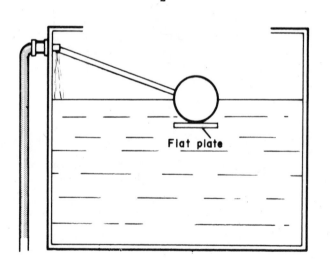

Flat plate

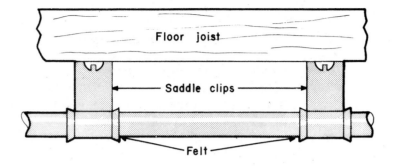

Floor joist

Saddle clips

Felt

3

SOUND INSULATION
1. Pipe 1in (25mm) 'sleeving' holes through structure.
2. Insulating pipework from structure.
3. Method of preventing self-induced water hammer.
4. Ball valve with silencer tube.

4

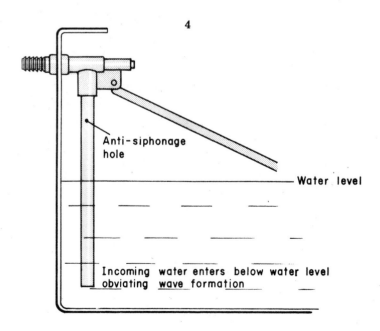

Anti-siphonage
hole

Water level

Incoming water enters below water level
obviating wave formation

A simple expedient but one that saves much on noise and shattered nerves is to fit door stops. Doors slamming are probably one of the most irritating and noisy sounds in the house.

Noisy plumbing can also provide much irritation, especially at night when sound — almost unnoticed during the day — is magnified. In former times traditional materials such as lead pipe and masonry walls tended to absorb some of these sounds. Modern copper tubing, being rigid and having thinner walls, transmits sound more easily.

In the planning stage it is a good idea to isolate the WC and bathroom from living area. WC pans and flushing cisterns are better placed on external walls rather than partition walls. Pieces of felt or rubbers will help to isolate them from the structure. Building Regulations now require internal soil pipes and where these pass through a room they should be boxed in and the box filled with loose-fill or polystyrene.

Pipes should be fixed at regular intervals with pipe clips and can be sleeved in pipe-wrap at points of friction and expansion through the structure of the building.

A cistern can be placed on a bed of polystyrene or felt. The filling of a noisy WC cistern can be reduced by fitting a short length of pipe from the ball-valve to the bottom of the cistern. The incoming water then enters below the level of the water instead of dropping on to it.

Simple, or near-simple remedies all — here are suggestions as to how you might make your home the quiet peaceful refuge it should be. Obviously it will be seen that many of the suggestions are the same or vary only slightly from requirements for thermal insulation. Thus, in deciding to insulate to whatever degree, one should be thinking of physical and mental insulation as well.

In the field of sound insulation one cannot expect miracles. Obtrusive sounds will still stop us dead in our tracks and obliterate the vital sentence in a TV play, but we can, nevertheless, do much to help make our homes quieter places.